EVERYBODY FARTS

written by
Haily Piccinin-Laferriere

DEDICATION

TO BRADLEY, ALLISON, BRANDON, AND OUR PETS

Let's talk about something important: passing gas! We all do it, and guess what?

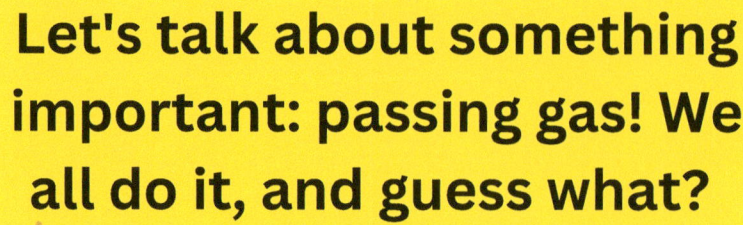

Bradley farts when he is playing video games.

Allison farts when she is eating dessert.

Harriet farts when she is eating treats.

Mel farts when he is napping

Mom farts when she is meditating.

Dad farts when he is riding his bike.

It's completely natural. Our bodies use it as a way to stay healthy. Do not feel embarrassed because, though it might be a bit smelly, it is a big relief once it's out! In fact, did you know that even plants and animals let out gases? It's true! From the tiniest insects to the mightiest elephants, and yes, even the lovely flowers in your garden. Everyone participates in this natural process.

So, don't worry! You're not alone in this 'toot-full' adventure.

We all fart, and thats O.K.!

Toot Toot!

Hey, who did that?

Silly kitty!

Meow

I love to love

I love to love
I am loved

I love to love

I am loved

I matter

I love to love
I am loved
I matter
I am kind

I love to love

I am loved

I matter

I am kind

I can make a difference

About the author:
Haily Piccinin-Laferriere, born in the lovely month of December, to a loving family in Massachusetts, has a passion for educational writing and enjoys writing children's books. With a fun and humorous writing style, Haily explores putting a playful twist on ways to teach children different facts about self and living a happy life. Beyond her literary pursuits, Haily can be found hanging out and spending time with her family, making loving memories together. Just as her stories put a smile on her family's faces, her intent with publishing her stories is to bring joy and smiles to all her readers.